Escape to the Country

LIVING ON THE FARM

Ben Ashby

Lannoo

CONTENTS

Walton Creek
FARM

Hemmed in by cornfields and emerald grasses blowing in the wind, dusty two-lane roads wind and weave across the land with reckless abandon and little memory of the property lines that once defined their curves. Here and there they reach a small town, sleepy and bucolic – a timeless refuge from the modern world and the megacities that seem to have dominated the cultural focus across the globe in recent years.

This way of life was once seen as lesser, as we were told to seek the big city lights and the perceived high culture of urban density and skyscrapers. But as digital connectedness has become an integral part of modern life, city living is no longer required for what many would consider an enriched life. Over the past few years we have seen countless city-battered millennials find refuge – and an equally enriched life – in one of these small towns set among lush fields and rolling hills. Here, we set out to visit these rural havens, taking in their scenery, traditions and culture, which have never ceased to appeal to so many.

I grew up on an old farm in rural western Kentucky. My family had been on that plot of land since the 1780s, and I often say the waters of Waltons Creek flow through my veins. At the end of the 18th century, we came across the mountains from the newly established United States in search of uncharted land. We settled along the banks of that tiny creek just down from the convergence of the Rough and Green Rivers, where the soil was filled with clay and the forests were thick. That valley has been home to the family ever since. Generations of Ashbys have lived there among the goldenrod and scrubby cedar trees.

Today, the valley is the community of Waltons Creek, population 50 – that is, if you count everyone up and down the old, winding one-lane chip and seal roads and the highway that cuts diagonally through it all. An old white church and a cemetery sit in the middle of a knoll, but the one-room school is long gone. The valley is quiet and calm, filled with the kind of folks they refer to as 'salt of the earth' or 'good people'. Growing up, I didn't know there was any other kind of life; there was no glimpse of the soon-to-be digital globalized world. Along with the tiny town of 300 a mile or so down the road, it was the kind of community that reminisced fondly about the sepia-toned life before the big-box stores came to the area.

When I was four or five, I learned that the house my grandparents lived in, the house my dad grew up in, the house my great-great-grandparents built, was handed down from generation to generation in the family and one day would be mine. As an only child, I grew to love that house, even insisting that it must be saved after my dad's endless comments about it needing to be torn down. I knew this place was connected to me. To me this house was the manifestation of the family that came before.

I grew up two doors down; our family had three houses in a row along the one-lane road. My Aunt June lived between us – her house is where I learned to cook, along with so many of the things I love today. In the gardens with my aunt and grandfather, the gardens I now write about, I learned the proper times to plant, to fertilize, to hoe, to weed, to pick, and to wait. I grew up in the pumpkin patches and the watermelon fields. I grew up on that land, immensely proud of the land and its history. Over the years my life changed and grew, family came and went, but my love for that piece of land continued.

In the spring of 2020 the old farmstead, the one my great-great-grandparents had built, became mine. My grandmother moved to the retirement home in town, and the deed was drafted for me to take over this place I viewed as sacred. The trends of the world had started to shift from the idea that life was best lived in urban places, and the idea of coming home, to the farm was put into motion.

I know countless people who have moved from the urban centers to small towns. We have seen a large exodus in the past decade thanks to the spread of high-speed internet to rural areas and the realization that many jobs once viewed as exclusively office-based can now be done remotely. But the trend had been gaining momentum for much longer, as fatigue of the urban areas and their fast-faced life often leads to burnout. Many people have been left craving a slower pace of life; tangible, in-person, community; and the ability to feel they are creating meaningful change. The luxury of affordable land, larger homes and lower costs of living have been increasingly palatable to many city dwellers all across the globe. We now also see a generation seeking to create goods with their hands and to create a life centered around nature rather than concrete grids.

As trends such as farm-to-table, organic cooking, and a focus on environmental issues have gained traction in recent years, the work of farmers and craftspeople has grown more visible. The idea of knowing where your food comes from has been embraced across the globe within all levels of socioeconomics. A desire to grow their own food has spurred many to leave the cities and to settle on their own farmsteads. As we see more and more people embrace these rural lifestyles, it is very clear that this is no longer a trend but a change that will gain momentum.

Although this lifestyle offers many possibilities, it is definitely not for everyone. Necessary concessions must be made when leaving areas typically considered the epicenters for culture and amenities. The rewards of growing your own food, having your own space, and being part of a small community must outweigh the negatives of less centralized access to shopping, entertainment and other modern conveniences.

When considering a move from the cities to the country, you must be ready to give up on some of the luxuries (expect spotty internet and cell phone service) and ready to get your hands dirty, quite literally. This living-off-the-land, small town–focused lifestyle may seem romantic, especially when it pops up on social media feeds and in magazines, but in reality there are as many downsides to it as there would be downsides to living in an urban area. Yet it is a lifestyle change that, if carefully planned, can be immensely rewarding.

My journey from the city to the country was one I had been quietly and subconsciously planning for most of my life. I had long dreamed of turning the old homeplace into something that fostered community. I dreamed of the garden, the patches, the orchard, the yard, and the barn lot once again returning to life. I reminisced of an idealized bucolic place that existed somewhere between the reality of my childhood and the memories my family shared of growing up on this same plot of land.

As I inherited the house and made my move from just across the Hudson River from New York City to the 120-year-old farmhouse in rural western Kentucky, the world was changing. The countryside provided room and air to breathe, and a place to roam in the quiet, away from a noisy world.

My years of collecting *Martha Stewart Living* and *Country Living*, my love of Alissa Hessler's *Urban Exodus* blog, and my own collections of stories from my hometown prepared me for a move. Granted, I was basically moving back to where I had started, two doors down from my childhood home. I had family, friends, and decades' worth of stuff stored in barns, garages, and buildings working in my favor. This move and transition has been made endlessly easier thanks to my dad knowing this land better than I do, frequent visits from my 83-year-old great-aunt who grew up in this house, and the community that joined me in bringing to life the events we host at the farmhouse. While I wouldn't recommend hosting a dinner for 50 in your first year of owning a rural homestead, I do have one, number-one tip: become a part of the community. Life is truly best lived in community, especially out in the rural parts.

BEN ASHBY

Spring

Life slowly returns on the farm.

As late February turns to early March, the daffodils begin to make their annual march from the frozen ground toward the warmth of the sun. Green shoots peek through, the first sign that spring will soon come. As the days gradually get longer, the crocuses join in. Color and life peppering the sleeping land.

For those with an especially thrifty green thumb, seeds are started in a humidity-laden greenhouse bought at the local Rural King or Tractor Supply Company, which stand out against the snow-covered land during the inevitable early-spring snowstorm. It is important to remember this is the most unexpected of seasons: some years the march of the daffodils begins in mid-February, some years it's days after the start of March. The best-made plans are often for naught as early growth and the hope of a warm spring is lost to an unexpected freeze.

Once spring comes to life, though, it becomes a symphony of colors as the bulbs, bushes, and trees join the triumphant march. The garden slowly warms as the fallow ground is broken and dried by the sun. Rows are marked; seeds for radish, snow peas, kale, and other root vegetables are sown, in hopes of garden-fresh dinners in the months to come.

In the house, the winter gives way to fresh daffodils, jonquils, and flowers and branches brought into the house, eggs dyed for Easter, and lettuce, radish, green onion, spinach, and kale brought in from the garden. A good spring clean and the home is ready for future guests; the swings rehung on the porch and in the yard. The feeling of life at the farm has truly returned.

Connecting to the Outdoors

Saratoga Springs, New York

When driving up to the upstate New York home of former Brooklyn residents Bethany Bowyer Khan and husband Jamil Khan, you start to question if this is a very old home or a very new build. The lines are just a bit too straight and the angles too sharp, but the style instantly feels like a home that could be centuries old.

Spring arrives a bit later here in Saratoga Springs. The daffodils take their time to bloom, long after the blooms in the southern states are already spent, but that makes the spring here all the sweeter.

Peas

Fueled by a desire to retreat to a slower pace of life, Bethany and Jamil left Brooklyn in 2020 with their young children in tow. Jamil, a Brooklyn native, and Bethany, with rural roots in northeast Ohio, were ready to live a more intentional lifestyle closely connected with the outdoors. Bethany left her fast-paced career in real estate and economic development to launch the sustainable design and lifestyle company Arcadian Revival, inspired by her own country upbringing and her home renovation. Jamil, who works in renewable energy (the couple met when working in energy policy for the City of New York), is also adjusting to the luxury of working from home. They are embracing their time together and connecting with their new property, from watering the garden to talking about the wildlife they've spotted on their doorstep: turkeys, owls, foxes, songbirds, and the occasional neighborhood bear.

A Storybook-style Cottage

Pleasant View, Tennessee

Somewhere in the rolling hills outside Nashville, Tennessee is the unexpected home of Christie Jones Ray. Christie, an artist and writer, moved to this home several years ago after living in the suburbs of Nashville for many years. Desiring more space and less traffic, Christie and her family purchased a new build in a subdivision in a town that to many feels straight out of Southern literature.

The town, still home to an old-fashioned feed and grain store and annual Christmas parade was an ideal place, but the home they purchased felt boxy, barren, and new, the opposite of Christie's signature storybook cottage style. The family set out to transform the new-build house into a cottage-inspired home. Christie, along with her husband Tim, spent several years searching the South for cottage and farmhouse-style antiques that would give the house the much needed charm. "The architectural pieces of riverboat gothic, ancient corbels, cozy quilts, and bouquets of homegrown roses and hydrangea alongside chippy cupboards and benches were the elements that created the cozy cottage abode we have grown to love," Christie explains.

Over the years a cottage within those dry-wall walls came to life. Inspired by Beatrix Potter, Tasha Tudor, and Rachel Ashwell the house, to the outside viewer, felt like a place that had been a home for decades.

Eliza Visits
Martha's Vineyard

"My upstairs studio is a retreat with its storybook sofa, wonky pink cupboards, and twinkle lights strung through the piece of gingerbread trim that hangs over the old oak table," says Christie. "A jelly cupboard filled with old books is the library in the living room. Really Old Bear's bed and rocker are tucked snuggly at its feet beneath the cabbage roses of the cottagey drapes." The best part about the embellishments the family have added to the home is that they're all movable and removable and will easily transform their next house into a home, if ever needed.

Kim's
GREENAWAY
COLORING BOOK
Eliza the Mouse

The Old Maple Wood Farm

Hudson, Ohio

Gina and Brian Bishop live at Maplewood Farm, on a three-and-a-half-acre plot that was once part of a 300-acre maple tree farm. Today only one of the original farm building remains, but it is right within walking distance of the darling historic town of Hudson, Ohio, just east of Cleveland. This area was settled by New Englanders who were promised land in payment for their service in the Civil War; it was also a well-known stop on the Underground Railroad.

A Federal-style wooden structure similar to those built in New England, the farm was once home to the only doctor in the Western Reserve. Later, the son of the original owners won a gold medal at the World's Fair in Chicago for his invention of the Maple Syrup Evaporator. A cheaper alternative to sugar at the time, maple sugar was produced in the farm's many barns. Now the last remaining barn, a century-old building located just behind the house and probably originally used to store carriages, is home to Gina's workspace and sought-after barn sales.

Many of the original windows and floors remain in the main house, which is actually three buildings in one: the oldest, central part dates from 1824, while the front and back extensions were added in 1850 and 1950 respectively. The front has larger rooms, as that particular part of the house was used as an infirmary. The office/den that Brian uses now was the doctor's office where patients would ring a bell that is still there. It is Brian's favorite feature in the home.

PICK ME UP

Gina has decorated in a way that honors the age and bones of the historic home. "I have added whimsy and unexpected items to each room, and have placed the items in ways that wink to just me," says Gina says of her unique style. "I try to not take myself so seriously and love color and mixing patterns so I just go for it. I do not really adhere to other people's rules when it comes to decorating and just do what feels right for me. People describe my home as happy, and I take that as the highest compliment."

The sunroom is one of Gina's favorite rooms and the family sits there for morning coffee and after dinner. It looks onto a wisteria arbor, and it is beautiful in all seasons. Gina has collected all of the needlepoint artwork and paintings in that room, as well as pillows made by friends and some pieces from her Homegirl home decor line. "I have so many quilts so I layer them; I feel like that's a granny's room but for some reason it feels really nice for me to be surrounded by all of those cute things," she says. "As for the office, again I just mixed what felt right. The day bed acts as guest room when needed and the fireplace never stops come winter." In the living room, Gina wanted to tone down the whimsy with leather sofas and other heavier furnishings, but "even if I try to be serious in execution of design it just never feels like me until I add that smile element," she laughs.

Since moving to their home nearly two decades ago, Gina, Brian, and their two daughters, Lila and Edie, have created a colorful yet cozy home. Gina's reputation for being the Mother of Barn Sales has made her a voice for the Americana style within much of the United States: "I could not love my quirky, whimsy house any more than I do and am happy to call it my home."

DECORATE
HOME DESIGN IDEAS
DETROIT

TAEH8
BALL M2

"PEOPLE DESCRIBE MY HOME AS HAPPY AND I TAKE THAT AS THE HIGHEST COMPLIMENT."

——— *Gina Bishop*

Bloomsbury Farm

Smyrna, Tennessee

Nestled in the rolling hills of middle Tennessee is Bloomsbury Farm. This three-generation business, run by Lauren Palmer alongside her parents and daughter, has become an idyllic escape from the near-constant construction and growth of nearby Nashville, a booming city that has allowed Lauren's vision for the farm to flourish in recent years.

BLOOMSBURY
ORGANIC FARM

Lauren graduated from college with a social work degree and worked in hospitals, but never felt entirely fulfilled during those years. She decided to intern on a farm that embraced organic farming, and says it was the hardest work she'd ever done, yet was absolutely the most satisfying. When she was a child, her parents had moved their family to a plot that is now part of the farmland she oversees. Both her parents (her dad comes from the area and was raised on a dairy farm nearby) encouraged her desire to farm the land, and soon they became her first, and cheapest workers. Over a decade later Lauren still says that as soon as she hit the dirt she knew she'd found her calling.

Today the farm has grown and is a year-round operation. Much of what you see on the farm began at least two seasons prior as Lauren plans well ahead to ensure her spring Community Supported Agriculture (CSA) kicks off each year without a hitch. Spring is a welcome season for the farmers here at Bloomsbury Farm. Warmer temperatures and longer days mean the 32 weeks of the growing season are soon ahead.

"In winter, our daily paces slow a bit, allowing us to take stock of the growing seasons that year and calibrate for the seasons ahead. We're constantly looking for ways to grow, and in spring we start to see that work take root," explains Lauren. "Spring allows us to reconnect with the community in a bigger way." While they attend the local farmers' market year-round, spring brings the CSA community out and back onto the farm. Lauren says that "it's like welcoming family home after time apart, which feels appropriate during the time of rebirth and growth that springtime brings."

For Lauren springtime is filled with sensory threads, from the sweet taste of the first ripe strawberries eaten out in the field to the chirping of baby chicks and the small buds that decorate the branches of the pear trees.

The farm is a Good Agricultural Practices and organic-certified and grows over a hundred varieties of vegetables, fruits, herbs, sprouts, and wheatgrass. A flock of chickens and gaggle of geese provide fresh eggs, and in recent years they've also started growing hemp. Through their on-farm market days, local farmers' markets, robust CSA programs, as well as wholesale channels and local restaurants, they put their goods into hundreds of local hands each year.

Over the years Bloomsbury has added summer camps (with a guarantee to send your kids home dirty or your money back) and a nature-based school. This last component—education—when combined with farming, feels full-circle for Lauren. The farm is a business, of course, but it is more importantly a place to nourish, educate, and connect.

"I'll never forget my first farmers' market in East Nashville," recalls Lauren. "Incredibly nervous, I wouldn't have done it if not for the support of a friend who said to show up with the produce and he'd have a table and tent ready. With a handmade poster board sign (which I still have) saying 'Grown with Love,' I jumped in and fell in love." Lauren continues: "Each year (and sometimes day) we have bumps in the road, but when life gives you an excess of tomatoes, you turn it into Bloody Mary mix."

The most important thing she's learned in growing the farm is to listen: "Listen to your community and customers ... to the people who've been with you the longest. Listen to what they want and need, then deliver." From the first farmers' market, Lauren fell in love with the connection you get to those you feed. "As farmers, we need to be face-to-face with our customers to listen and learn what they need. Our family motto has always been 'bloom where you were planted.' Bloomsbury is more than a place to grow, it's a place to gather, and I believe that good community, like good food, doesn't just happen; it is something that's planted, nurtured, and shared."

"WE'RE CONSTANTLY LOOKING FOR WAYS TO GROW, AND IN SPRING WE START TO SEE THAT WORK TAKE ROOT ... SPRING ALLOWS US TO RECONNECT WITH THE COMMUNITY IN A BIGGER WAY."

——— *Lauren Palmer*

In Spring

The vegetable garden
is a rural tradition as old as the homestead itself. Passed down through the generations, in many ways the methods used in the household garden are today the same as they were hundreds of years ago. Even if almost every community now has farmers' markets, and locally grown produce and goods are readily available, for many the desire to grow their own is as strong as it ever was.

START YOUR GARDEN

Gardening is truly a year-round activity, but for those who buy plants from a nursery or a greenhouse like I do, it begins in early spring. For those who start their plants from seed, the season will typically begin earlier, in January, with the annual arrival of the Burpee and Burgess seed catalogs—an old-fashioned tradition that lives on even in the modern age of internet ordering.

If you are planning on starting your garden from seed, it is important to look at germination and growth rates as these will vary for each variety of plants. You do not want to start seeds too early and run the risk of the plants growing too quickly indoors. It is important to remember that when selecting where to start your seeds you must have an abundance of light and a warm environment. A greenhouse is ideal, but for the beginner gardener a sunny window or the use of grow lights will work just as well.

BREAKING THE GARDEN

You want to break your garden early in the spring, as this will help eradicate the weeds as well as provide the soil a jumpstart on warming. For most crops you will want to begin planting after the threat of frost or freeze has ended. Early spring crops like greens, radishes, or peas can be grown using cold frames or by surrounding them by straw or other bedding material.

The process of breaking the garden is the act of breaking apart the soil. You will have to plow the garden, going deep into the ground and flipping the soil. This will dry for several days before a disk creates a well-worked soil that is level and has few large dirt clots. If you are only breaking a small area of ground, a tiller may be used. This will likely require the use of hand tools to further break down the soil and to remove any detritus from the previous season. Once the ground is well worked, you are ready to begin planting.

PLANTING

Your rows should run north to south, in order to create even sunlight distribution throughout the day. When running rows east to west, taller crops will oftentimes shade out shorter crops in close proximity. When laying out your garden, keep in mind what will be planted where. Traditional rows are 18–24 inches apart. I find it helpful to know the width of your tiller and to ensure that it will fit down your rows as your plants grow. A tiller will greatly cut down on time and energy spent with weeding.

BAKER CREEK HEIRLOOM SEEDS
MANSFIELD ★ MISSOURI 65704
KILIMANJARO
POPPY
BLACK PEONY
BAKER CREEK HEIRLOOM SEEDS
MANSFIELD ★ MISSOURI 65704
CALENDULA
SNOW PRINCESS
BAKER CREEK HEIRLOOM SEEDS
MANSFIELD ★ MISSOURI 65704
SALPIGLOSSIS
CAFE AU LAIT
NASTURTIUM
ALASKA MIX

I
FOR YOU

SHOP TOUR

Lonesome Pine Home

Texas

"I have this feeling deep in my bones that my town has the potential to be something great and I want to be a part of that," says Hemphill native and resident Samuel Melton. In 2017 Samuel bought a dilapidated 1920 men's clothing store on the square of this small east Texas town. He first established his shop, Lonesome Pine Home, in a rented space on the opposite side of the town square, before moving in after extension renovations.

"I moved from the big city [Dallas] with $100 in my account back to my hometown to start a shop that may or may not have worked. I believe firmly in doing the dreams you have before it's too late, and that's what I'm doing now," explains Samuel. His vision for the store is the perfect mix of old meets new, with exposed plaster walls and a focal wall of old tin ceiling tiles that greets you as you walk in. "I grew up in this life of antiques and vintage, it is in my blood," says Samuel. "I watched my parents hunt for those treasures all around the South as a kid. I grew into finding what I loved, which is furniture and things that have their own stories."

Samuel's largest project in this small town has been remodeling his great grandparents' little house near the Sabine National Forest. "It has been an act of love to my heritage." Old paneling has been replaced with large planks of 12-inch wood from his grandfather's furniture store. The kitchen is lined

I
FOR YOU

E
T
X
O
R
B
U
S
T

with open shelving in blond hickory he had planed from a tree on his land. "I used found lighting and mixed it with hand-forged Texas-made lights to give artisan touches. I also laid local handmade tiles from Thrō Ceramics on my entryway floor. The little house is not much, but it's mine. This is something I couldn't create when I was in the city."

"We are a town of people looking for a life different from most. I like the term *urban exodus* to mean 'make your life on your terms in a small town hidden in lush pine forest.' We sit on one of the largest lakes in Texas dividing us from Louisiana. With a small town comes small businesses that I love to support: meat markets, quilt shops, new wineries, and many more places to explore here. I try to tell anyone who walks in the shop about this hidden gem I call home. My town is a underdog story waiting to be told with many flocking here finally finding old buildings and homes waiting to be restored. I have that feeling in my bones of 'something big is coming'."

LOCALS

"WE ARE A TOWN OF PEOPLE LOOKING FOR A LIFE DIFFERENT FROM MOST."

——————— *Samuel Melton*

RECIPE

Chocolate Banana Bread

INGREDIENTS

2 sticks (225g) + 3 tbsp butter
3 cups (385g) flour
2 tsp baking soda
2 tsp baking powder
½ tsp salt
3 super-ripe bananas, previously frozen and thawed
4 tsp lemon juice
3 tsp pure vanilla extract
2 cups (400g) sugar
2 large eggs
1 cup (170g) chocolate chips

This recipe is from my friend Jen O'Connor. Whenever you go to her house, she always has this bread freshly baked and under a cake dome on her kitchen counter. It is a dessert that can easily be served at breakfast. This recipe will make two 8–9-inch (20–22cm) cast-iron skillets or one Bundt pan. If using a Bundt pan, coat with baking spray, as this cake tends to stick.

1. Heat oven to 350°F (180°C). Drop 3 tbsp butter into the skillets and place in oven to melt (you want them very hot and bubbly before you pour the batter).

2. Whisk flour, baking soda, baking powder and salt in a separate bowl; set aside. In a stand mixer, beat the butter, thawed bananas, lemon juice, and vanilla. Add the sugar and beat until as creamy as possible (it will still have lumps). Add the eggs one at a time and beat through; add the chocolate chips and mix lightly.

3. Put half the batter in the center of the skillet or Bundt pan. Bake in the center of the oven for 1 hour or until a fork comes out clean. Cool on a wire rack before serving.

RECIPE

Peach Cornmeal Cobbler

INGREDIENTS

1 stick (110g) butter
1½ cups (300g) white sugar
1 cup (215g) brown sugar
1 tbsp cornstarch
4 cups (675g) sliced peaches (peeling is a preference)
½ cup (70g) self-rising flour
1 cup (180g) cornmeal
½ tsp salt
1 cup (225 g) buttermilk
cinnamon, to taste

Everyone in the South has a peach cobbler recipe they use. Most are that cuppa-cuppa-cuppa sort of thing that has like four ingredients and can be made just as easily with a can of fruit cocktail as it can with fresh peaches. There is a time and place for this sort of recipe, but during peach season you deserve better … and this is the recipe for that. Cornmeal and buttermilk are two of the most Southern ingredients and it truly surprises me that they aren't used with peaches more often. I add cinnamon and brown sugar to create a crispy sort of crust that has the taste of Cap'n Crunch.

1. Preheat oven to 375°F (190°C). While the oven is preheating, place the butter in a baking dish and melt.
2. Place one cup (200g) of the white sugar, half a cup (100g) of the brown sugar, the cornstarch and the sliced peaches into a pan. Bring to the boil, stirring continuously.
3. In a mixing bowl, combine the remaining sugar, flour, cornmeal, and salt with the buttermilk. Combine only until wet; do not overstir.
4. Pour the mixture into a baking dish; do not stir. Pour the peach mixture on top; do not stir. Sprinkle some cinnamon on top. Bake for about 30 minutes, until golden and bubbly.

ASHBY

Summer

IN
BOTTLES

The summer season on the farm

is when the homestead revels in its glory. The narratives of great rural literature come to life in a timeless dance that transcends time and place to create a reality that feels like a romanticized tale of the past. Cicadas lament the passage of time as their chirping drowns everything nearby. The decades-old maple trees' deep, dark shade shrouds the yard and the house with a coolness, warding off the humidity that clings to every surface left in the sun.

The garden is in full bloom and grows with each day. Rows of tomatoes, corn, Kentucky pole beans, and okra are planted, hills for the squash and cucumbers are created nearby, and the tables near the back of the house and cistern are covered in a daily harvest of squash, tomatoes, and cucumbers. Neighbors, visitors, and the UPS man go home with sacks of produce that truly feels endless.

Sweet tea is served to everyone in an attempt to compensate for the summer heat. Air conditioners in the upstairs bedrooms thrum constantly. Beds wear their summer-weight clothes and every menu calls for lighter garden fresh fare. Life is truly slow and sweet in the summer. The sun rises long before the first of the rooster crows from a neighbor's yard and sets only as bedtime calls. The days are filled with long work and time spent in the garden and the yard, but the season provides its prolific rewards with every passing day.

A Rural Oasis in the City

Boston, Massachusetts

The dream of a life out on the frontier, on a farm, or in a tiny town is one that someone might have, but because of career limitations, it just isn't feasible for everyone. That however doesn't mean one can't have a home that feels like a rural oasis in the city. One man that took his rural dream to the city is Philip Saul. He knows that natural elements, vintage and antique finds, and treasures collected from small town adventures can easily create a space that feels fit for any small town. Somewhere in the heart of Boston, Philip has created an apartment that would feel right at home among the farmhouse and the cottages.

Philip, the owner of Boston-based Sault New England lives within walking distance of his main shop. For the past decade he has brought quality New England goods and clothing to the shoppers of the city, but for much longer Philip has been on the hunt for the perfect finds and antiques to create his own home.

"ONE MAN
CAN MAKE A DIFFERENCE
AND EVERY MAN
SHOULD TRY"

As an avid collector, Philip makes annual trips to Brimfield Antiques Show, one of the largest antique and vintage markets in the United States. Located west of Boston, the tiny town of Brimfield comes to life multiple times a year as hundreds of antique vendors present their most recent curation of goods collected from across the United States and the globe. Philip has developed an eye for spotting the must-have pieces that complement the style of both his shop and his home. Blending classic New England style with a nautical feel, his finds seem to embody the endless stories of past times.

NAVIGATION

ARTUS FULL-VUE
SPARTUS CAMERA CO
CHICAGO, ILL.
MADE IN U.S.A.
DECORATE

A House That Tells Stories

Mount Pleasant, Iowa

The fields of Iowa have a way of growing without limits, stretching to the horizon and beyond. East to west, north to south, the rows of corn and beans feel eternal. Life out there on what was once the plains and prairie feels timeless. The farms transition from one generation to the next as the combines and machinery get bigger and the passage of time is marked by ever increasing horsepower and hybrid varieties of grain that will feed the world. Kaleb Wyse grew up right here in the middle of it all, on a family farm that stretches to the horizon and is bisected by straight-as-a-crow-flies roads.

For Kaleb, home is a house that was built in the 1880s. Its original Victorian details were removed in the mid-1900s, and today the house has more of an American four-square appeal. His grandparents purchased the home and farm as a place for his parents to move into once they married, before swapping houses and moving in themselves. "I feel a sense of belonging living in homes where past generations of family have lived and worked," says Kaleb, who purchased the home from his grandparents when they chose to move into something smaller.

"I restored the main floor with some original details. I matched all the woodwork to the original second-floor pieces. I opened some doorways to make the flow better and updated the kitchen, but kept the heart of an old house in the design. Cresting a story of what could have been there all along." To almost anyone else this is the middle of nowhere, but for Kaleb this is the place that feels most like home. His father passed away in 2004 and he says he feels connected to what his father loved and where he loved to work on the farm.

"I have collecting in my blood. I am constantly finding pieces that tell the story of what I love. I look for items that are timeless. They could have been in the house originally and been used daily, like yellow-ware pottery. I love to create a home that feels like it has stories to tell, like the items in the home have lived with the house. I have no rhyme or reason but choose what makes me happy. If I fall in love with it, I know I will find a place for it."

THE WAY TO COOK
EVERYONE
Love real food
FOOD IS THE SOLUTION
ROSE'S BAKING BASICS
Whole Food Cooking Every Day
COOL BEANS

An Artist's Farmhouse

Water Valley, Mississippi

The state of Mississippi has been long known as a very Southern state, one that is filled with vast farms, small towns, historic charm—and, of course, the mighty Mississippi River. Water Valley, the town Jonathan Kent Adams and Blake Summers call home, is located just south of Oxford and was once a major producer of watermelons. Today it is a much sleepier place, which suits Jonathan and Blake well. Their home sits just a few blocks off the center of town, a pleasant walk from restaurants, shops, Jonathan's studio, and the folks that make this town feel like a bit of a liberal refuge in an otherwise staunchly conservative state.

"Our style is whatever we like," says Jonathan. "We are a bit all over the place: Blake is usually drawn to darks, earth tones, ornate, and brass; I am more modern, playful, and bright. What unites our style is our love for the pieces we both pick out." The pair bought their house, which was built around 1910, from the artist Coulter Fussell a few years ago. "I am hoping a little of her magic is still there for me to tap into." She moved a street over and still makes work in Water Valley. Since taking ownership they have slowly made subtle changes to the house to make it work for their ever-active lifestyle. Each year brings an array of projects for the house and property. Their favorite part of the house is how it sits on a hill and for the evening golden hour light floods in each evening.

"I am not sure a lot about the history of the house. Our neighbor used to play in our yard and will tell me places where sheds or old walkways used to be. I know there was a small addition in the 1980s along with the vinyl siding. The addition is actually

FLOUR
HUSKY

where our roof still leaks. Tackling that will be one of our future projects," says Jonathan. A well-respected artist whose work has been exhibited and shown in shows and galleries across the United States, he has a studio a few blocks from the house.

Most items in their house were found at thrift and antique stores. Shops like The Depot, B's Antiques, Helping Hands, Sugar Magnolia, and the Mustard Seed in Oxford have been a constant source of new additions to the home. Blake frequently takes trips after work to see what's new or the pair will take little trips together to shop for new finds. Jonathan advises to check your favorite thrift and antique stores frequently for changing inventory and for pieces you may have originally overlooked.

When asked why they decided to move to small-town Mississippi, Jonathan answers: "Because it's cheap, and because we want other queer people in Mississippi to know that other LGBTQ people are here! Here we can afford to be creative without having to spend forty-plus hours a week stuck in an awful job."

The Land, Trees, and the Pond

Catskills, New York

The move from urban areas to country life isn't a new trend. We watched Oliver Wendell Douglas and his reluctant wife Lisa move from New York City to wherever Hooterville is in the 1960s sitcom *Green Acres*, and the narrative has continued to play out on television and movie screens since. But for many that desire to move to the country isn't about following a trend or some cinematic inspiration. Long Island-native Jill Weiner made her move from the city to the country of the western Catskills in the mid-1990s. She grew up with close access to New York City, but was from one of those small Long Island towns where people would still occasionally pay her father, an accountant, with homegrown tomatoes.

After her childhood in the suburbs and college in Ithaca, Jill moved to Manhattan but by the 1990s she was ready for another change. "The city in the 1970s, when I would visit as a child, was gritty and in the 1980s, when I lived there, nonstop fun, but by the mid-1990s I needed a nap! While in Hell's Kitchen I ended up falling in love with pottery and had a recurring dream that through a closet there was a backyard, a big outdoors, and I needed to find that piece of green in real life. So I headed northwest to find home. I found my destination in the back of *The New York Times Magazine*, between the ads for summer camps and Sansabelt pants. It was for a house that turned out not to be for me, but led me to Sullivan County, two hours from the west side and the center of what was the Borscht Belt."

Happy hearts and happy faces, Happy play in grassy places— That was how, in ancient
WHO INVITED YOU ?

"The very next weekend I went right back upstate and found my home, one turn down a very long and witchy driveway and I could exhale," Jill continues. "It was not the house, which was a mess, or the weedy and overgrown gardens. It wasn't the falling-down garage with the attached porch nor was it the in-need-of-a-paint-job-and-thorough-clean-out of a barn, but it was the land, the trees and the pond that had me smitten. And smitten I remain. I bought myself 60 acres and just recently acquired an adjoining 12 just to preserve it from sprawl."

Jill has kept busy since her arrival in the country: "It has been 26 years of pulling weeds, evicting varmints, home improvements, painting, repainting, installing mismatched galvanized steel siding, vintage porch posts and twisty sluice pipes holding up the roof over the new porch." Over time windows have been replaced, solar panels installed, carpets replaced by wooden floors. A painted black porch now overlooks her cutting garden, which is filled with annuals, perennials and surrounded by massive walls of hydrangeas. "I have evolved from potter to wedding and event florist," explains Jill. "It seemed a natural change and as I grew into my home and my business, the Catskills have become a wedding destination so I don't really have to, nor do I want to, travel far from home to work."

"Moving to the 'country' for me was a gentle transition," she says. "I dipped my toe in slowly. I did not know a person for a hundred miles from my new home; it was a true leap of faith. I slowly met people and found my 'tribe'—other creative, fun, community-minded, compassionate people. Some were local (born here with generations of born-here ancestors); others, like me, transplants who found their place. I learned early that it's important to listen, to participate, and to lend a hand when one can.

Her own word of advice for anyone tempted by the same journey? "The biggest error that people make when they move to a new community is to try to 'fix' things that they think are broken or to assume that their new neighbors are 'less' than them in any way," she says. "I have seen it. I have been here for long enough to have been on the receiving end of it. It is not an endearing quality. Be kind to your neighbors, you may not agree on some big important issues, but you will be there to pick up their pieces and they will be there to pick up yours in all sizes of emergencies." And another very important piece of advice: "Do not dig up anything until you have a full year of watching what is growing. I have had to provide interventive and calming canoe rides for new neighbors who might have taken the shovel that their partner used to move some peonies and used it nefariously."

PHYSICAL AND POLITICAL
NEW YORK
NEW YORK
CATSKILL
MOUNTAINS
Syracuse
Utica
Albany
Binghamton
Buffalo
LAND ELEVATIONS AND SEA DEPTHS
LOW LANDS
DSON VALLEY
GAINST FRACK

"BE KIND TO YOUR NEIGHBORS, YOU MAY NOT AGREE ON SOME BIG IMPORTANT ISSUES, BUT YOU WILL BE THERE TO PICK UP THEIR PIECES AND THEY WILL BE THERE TO PICK UP YOURS IN ALL SIZES OF EMERGENCIES."

——— *Jill Weiner*

Farmstead Florals

Oxford, Mississippi

There is something really joyful about flower farms. Bright bursts of color filling row on row across the land, tall stems stretching toward the sky, waiting to be added to the most beautiful arrangements. A finite season of fleeting beauty, lasting only for a few days, but creating joy to fill entire spaces. Just outside Oxford, Mississippi, where the rolling hills start to rise up, is a flower farm, Farmstead Florals. Owned by Katherine Sharp, it brings seasonal beauty to cities all over the South.

"I've always loved being outside, sweating, and getting dirty, but that lifestyle became all the more rewarding when I added flowers to the mix! It is hard to believe that I am in my seventh season raising cut flowers for florists and groceries," says Katherine. "I came to farming naturally as my parents owned a vegetable farm in South Louisiana, but in summer of 2016 I decided to put my own spin on the family business and started growing flowers here in Oxford. The feeling of planting a seed, watering, weeding, pruning, and nourishing it as it grows bushy with blooms, coupled with the minor success that I had selling the stems, was enough to make me go out on my own and make Farmstead Florals my full-time job. Slow and steady has been the motto as I grew from three-quarters of an acre to 5 acres of open fields. From selling to just one of the local florists in town and the Saturday farmers' market, to now over 30 florists and groceries across three states, and two farmers' markets, plus our self-serve stand on Oxford Square.

"I LOVE PLAYING WITH THE DIFFERENT TEXTURES AND COLOR PALETTES AND WATCHING AS THE BOUQUETS CHANGE EACH WEEK FROM MAY TO OCTOBER AS DIFFERENT VARIETIES COME AND GO OUT OF SEASON."

Katherine Sharp

"Our meat and potatoes is selling wholesale to florists, but most people around town know us for the 'on your Oxford honor' self-serve flower stand we keep stocked 24/7 in the middle of downtown. Having not been an Oxford native, I was at first a little skeptical of whether an unmanned, glorified lemonade stand with a cash box drilled into the side selling mason jar bouquets would work or not. But having been rejected with an armful of flowers more than a handful of times, I remembered my business mantra for success, 'Some say yes, some say no, so what, who cares, who's next,' and decided to go for it! I was amazed at not only how well the stand did from the sales side, but also how quickly the community took to it, did right by it, and claimed it and my budding business as a new fixture of our small town. My nickname soon became 'the flower girl' and I loved it and was so happy to call Oxford my new home!"

"Never before did I think of myself as a creative person until I found flower arranging. I love playing with the different textures and color palettes and watching as the bouquets change each week from May to October as different varieties come and go out of season. In the winter months when there is little to nothing to do but wait for the rain to stop, I found myself itching for a new creative outlet. With the few flowers I had hung to dry as an experiment I made my first dried floral wreath. Pleased with the final product, we strung some lines of twine and turned an old out of use box refrigerator for vegetables into a drying room for flowers. Now in October when the frost comes and kills all of our flowers overnight we put away our mason jars and get to work making wreath rings out of branches around the farm and using floral wire to secure dried and faded blooms to our natural base."

"One of my favorite things about living and working on a farm is there is a never-ending to-do list and an endless amount of projects you can inspire. I feel so fortunate to have found my passion and been given the opportunities to thrive so early on."

In Summer

Cicadas fill the days
with their relentless chirping; crickets echo in harmony; birds fly through the air. An unbreakable humidity hangs and clings to everything the morning glory vines haven't crept up on. The garden in this summer season is in all of its lush and vibrant glory. Daily picking is required and the cornucopia of the season is continually overflowing. These are the months we prepare for the rest of the year. The implements, the hoes, the rakes, the clippers are in use, antique baskets line the table ready for filling. Flowers bloom and cascade around the rows of corn and beans; tomatoes stand tall, just begging to be picked and juiced.

The summer garden is one that truly does require daily tending, weeding and picking; tasks that everyone on hand must take up. To ensure an enduring and fruitful garden, the weeds must be removed swiftly and frequently, as it takes only a few days for unwanted weeds to take hold and choke out the plants you've already worked so hard to ensure will survive and thrive. Frequent picking of the ripe produce and the removal of damaged produce is a must to ensure the continued production. Plants such as tomatoes and okra really must be picked daily. It takes only a brief stay in the summer heat to overripen.

Each day in the summer garden will be long and endlessly fruitful. There will be plenty of friends and critters to keep you company, a cacophony of bugs, insects, birds ... and, if you're lucky, butterflies and bees. Like fertilizing and weeding, frequent watering (be it from sunny summer showers, late-night thunderstorms with their displays of electrifying light illuminating the midnight skies, or some sort of irrigation or watering routine) is essential, especially as the days grow to their longest and their hottest.

A garden is best enjoyed with friends; a bounty and a surplus are always the goal. While canning and preserving ensures a healthful diet in the long winter months, sharing fosters community—it feels good to be a neighborly neighbor. Grow with excess; gluttony doesn't exist in the garden when you're able to share. A basket of green beans, a bundle of beautiful golden ears of Peaches and Cream, our favorite variety of corn, or a sack of tomatoes, like Ouiser from *Steel Magnolias*, make a lovely gift that is truly from your own hands and land.

The best advice, beyond watering and weeding, is to keep crops off the ground as much as possible. It is important to surround your tomato plants with cages or stakes. I plant my tomatoes in long rows with stakes driven into the ground every two to three feet with twine interwoven between. As the tomato plants grow, train them up by adding twine and weaving the tender vines between. Keeping the tomato fruit off the ground helps prevent rot and helps keep ground vermin and insects away. Planting your green beans, butter beans, and lima beans with your corn will allow the vines to grow up the corn stalks. This once again keeps the produce off the ground but will make for a much more enjoyable bean-picking endeavor.

The summer garden is a time of beauty and change, of growth and abundance. The passage of time is a pleasure to follow and take note of. Simple country pleasures can oftentimes bring the greatest of annual joys. Watching the summer garden evolve is one of the most rewarding of pleasures.

Preparing much of the summer garden for the table couldn't be simpler. While enjoying a freshly picked tomato or peach right in the garden or orchard is a must-do tradition, things like green beans and sweetcorn are best enjoyed with minimal preparation.

Preparing corn on the table is a process that just takes minutes and can be done so many ways. I enjoy sweetcorn simply boiled: in a pot of water deep enough to submerge your ears, place your pre-shucked and brushed ears—use a stiff brush with a light movement of hand to remove the long fibers remaining after the shucks are removed—in the water. Cook for 3–5 minutes and remove from the water. Serve with good-quality butter and a light dusting of salt.

Green bean picking is a time-honored tradition in the South and in rural communities. Green beans require snapping and stringing (breaking the edible hulls into smaller pieces and removing the fibrous strings that run the length of the hull) and are oftentimes prepared en masse. Preserving beans in quart jars for the winter months is a simple process that has been a family pastime for years. Large buckets will be filled with beans, large pans put out, and the family will gather around the table, or on the porch in rocking chairs or swings, to spend hours breaking and stringing, talking, and reminiscing on times gone by.

For a simple way to bring green beans to your table, simply fill a heavy pot with water and bring to a boil. Add four to five cups of strung and snapped green beans, a scattering of salt, two to three strips of good-quality bacon, and a few pealed cloves of garlic to the pot. Bring to a rolling boil and cook for 20 minutes, stirring frequently. Reduce to a slow simmer and add water as needed. Cook for two to three hours or until the beans are tender and the house smells of the most delicious bacon, garlic, and green beans. Some recipes will tell you to cook your beans all day, but the desired tenderness truly is based on your own preference.

A Southern Supper

Waltons Creek, Kentucky

A generation or two ago, community dinners were common. Today, though, they seem to have fallen by the way of black-and-white television or the rotary phone. I'd often seen photos from old gatherings here on the farm at my great-great-grandmother's house across the highway from mine. There were dozens of family members there, an endless number of dishes, all homemade, and it was all out in front of her house. As a kid I would dream of what it must have been like to attend a dinner like that … I dreamed of one day hosting one on my own front lawn.

My friend Terri and I had been hosting events and workshops at my house for a couple of years when we decided it was time to bring this dream to life. We would host a dinner on the grounds at my house, everyone and anyone could come – they just had to bring a salad, side, or starter. We'd provide the sweet tea, fried chicken and the desserts. Our only rule was no talk of religion or politics at the table; we wanted it to truly feel like a place where anyone was welcome.

That first year we had friends come from near and far, a mix of 50 people, some coming from New York, Tennessee, and Georgia, and many coming from just down the street. Our second year we had nearly a hundred gather under the old oak tree between my house and barn. Folks from New York, Florida, Oregon, and Texas all came to be a part of our rural community. It is funny how in many ways these rural places know no strangers and everyone is welcome as long as you're willing to take part.

A
GOODTIME
was had by
ALL

If you're planning your own neighborhood supper—and you should—don't be overwhelmed. My friends Jen and Micah regularly host dinners for 200 or more at their house in Warwick, New York. The key is planning ahead and knowing how to delegate to those willing to help. Each year we have people show up early, eager to help set the table. Korey from the farmstead in town donates produce; Riley bakes the bread. The prep, like the dinner, becomes a communal activity.

The two days before and the day after the supper are more enchanting for Terri and me than the supper itself. We take pride in living in this small town and truly love getting to welcome our friends from elsewhere. The weekend begins with a steak dinner for those staying at my house; we take field trips to distilleries (Kentucky is known for its long history of bourbon production), to museums, to our favorite restaurants, and to simply see the small towns we love so dearly. We have found this slowed-down speed and relaxed cadence of the weekend to be something that is reflective of the slower pace of life in the country. There is no hustle and bustle of the city here, life moves at a slower speed and these summer nights are forever illuminated by the light of the fireflies.

DAILY
BUCKET

Terri and I have learned there is a novelty for many city folks as they explore the rural life, but we aren't trying to sell anyone on the idea of buying a house here and joining our Green Acres sort of life (although Christie Jones Ray did buy a house up the road). This dinner and our weekend each July celebrate why we love living out here. We are both from families that have been in this community for generations, and our pride of place runs as deep as the waters of the Green River to the west. Between the main course and the dessert at the supper we pass a basket to raise money for the annual Art Guild scholarship that is given to a high-school student each year.

French Toast

INGREDIENTS

Thickly sliced sourdough
or friendship bread
4 eggs (beaten)
1 can (400ml) sweetened condensed milk
½ cup (100g) brown sugar
1 tbsp pumpkin pie or apple pie spice
1 cup (240ml) milk

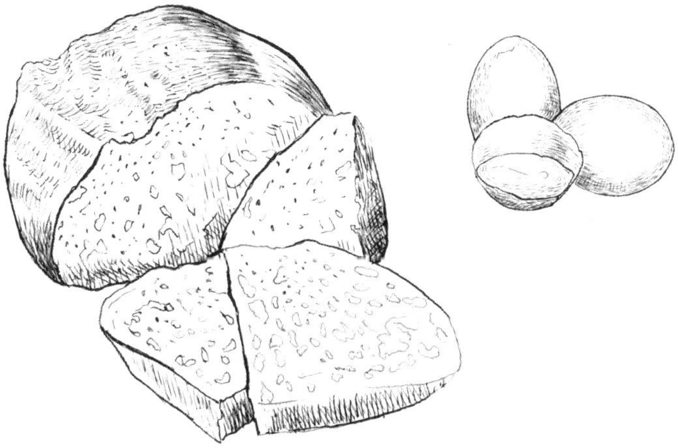

After the supper is over and the sun comes up, there are plenty of dishes to be cleaned and the tables and chairs must go back in the barn, to await the following year, but first we all have a French toast breakfast together, using leftover sourdough bread from the night before. The condensed milk and brown sugar will cook to give a delicious caramel flavor.

1. In a bowl, mix together all ingredients except the bread.
2. Heat an electric griddle or large skillet to medium heat.
3. Dip the sliced bread into the mixture until well coated.
4. Cook on a well-oiled griddle or skillet until golden brown and firm. Flip and repeat.
5. Serve with sliced bananas, a drizzle of caramel sauce, and maple syrup.

Old-fashioned Banana Pudding

INGREDIENTS

1 can (400ml) sweetened condensed milk
1½ cups (360ml) ice-cold water
1 3.4 oz box vanilla instant pudding mix
3 cups (690ml) heavy cream
4 cups (900g) sliced bananas
1 2 oz box Nilla wafers

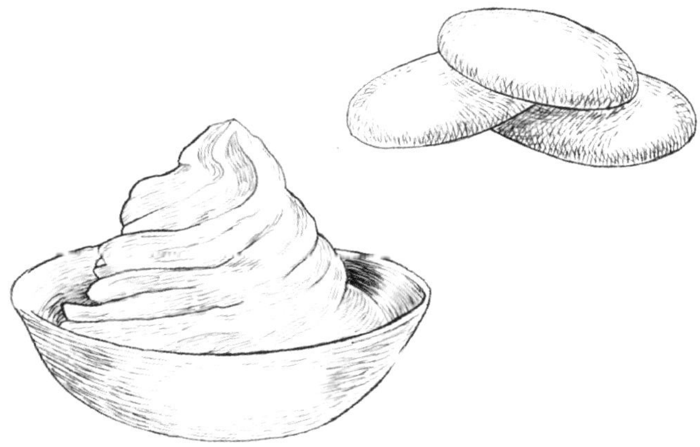

Growing up we had a very simple banana pudding recipe that I loved and could eat way too much of. Over the years I have altered that classic recipe to be a bit richer and more luxurious.

1. In a large bowl, beat together the sweetened condensed milk and water until well combined, add the pudding mix and beat well. Cover and refrigerate for 3–4 hours or overnight.

2. In a large bowl, whip the heavy cream until stiff peaks form. Gently fold the fresh whipped cream into the pudding mixture.

3. Arrange a third of the Nilla wafers to cover the bottom of a bowl or pan, overlapping if necessary.

4. Next, layer a third of the bananas and a third of the pudding mixture. Repeat twice more, garnishing with additional wafers or wafer crumbs on the top layer.

5. Cover tightly and allow to chill in the fridge for at least 4 hours—or up to 8 hours.

RECIPE

Fried Biscuits

INGREDIENTS

12 canned biscuits, cut in half
1 cup (200g) sugar
¼ cup (50g) brown sugar
1 tbsp pumpkin pie or apple pie spice

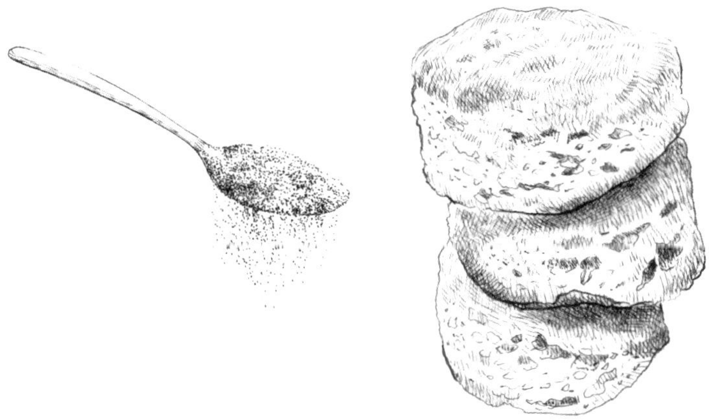

You have to trust me on this recipe. It almost seems too simple, but use the canned biscuits. I have made it with scratch-made biscuits and store-bought, and the store-bought biscuit dough simply works better. This is a recipe that can be made in excess for a party, but is best when the biscuits are served fresh and warm. These are especially delicious when served with apple butter.

1. In a deep-sided pan heat some cooking oil to medium heat. You'll know it is the appropriate temperature when a pinch of dough begins to sizzle when added to the oil.
2. Slowly add the halved biscuits to the oil. Turn once the bottoms are golden.
3. When both sides are golden, remove and drain on a paper towel-covered plate.
4. Mix together sugars and spices. Roll the fried biscuits in the mix and serve warm.

Fall

Fall has a way of creeping in here at the farm.

The deep greens of the corn, the trees, and the gardens slowly fade away, turning into shades of gold. The cicadas and frogs go quiet, and a constant dust hangs in the air in place of the humidity that lingered far too long.

It is no secret that fall is my favorite season. Of course, I love the beauty of each and every season: the warmth and the outdoors of summer; the rebirth of spring; even the calm and quiet of winter. Fall, however, is the season I look forward to most, with it spirit of thanksgiving, bountiful harvest, and stunning landscapes washed in reds and golds.

From the age of five to eighteen, I grew pumpkins each year to sell on my pumpkin stand, on my grandparent's driveway. In those early years I loved picking and arranging the pumpkins. My Aunt June and I would wrap one of the trees out front with dried corn stalks, creating our own oversized fodder shock to surround with pumpkins. Ghosts and giant plastic bag pumpkins were crafted to decorate the yard, and each year the homes along Chandle Loop came to life for the fall. I also loved gathering wild flowers from the creek banks and the fencerows. Vibrant goldenrods, tall grasses, and dried weeds were crafted into arrangements and tucked everywhere my parents would allow.

As I've gotten older, and now have my own house here on the farm, the season has taken on an even greater life of its own. Long before it is socially acceptable to begin decorating for the fall, I am begging the local pumpkin growers to sell me anything that might look remotely ready. Warnings of the pumpkins baking in the still-hot summer sun are readily ignored as I festoon the porches, the mantle, the counters, the dining table, and the yard with dozens of Musquée-de-Provence, my favorite variety of pumpkin.

Here in Kentucky, apple season on organic trees tends to be in mid-August. Old wooden ladders are taken from the garage, the heirloom baskets from Longaberger are taken out, and all my houseguests are responsible for gathering the apples from the trees and those on the ground. Cast-iron skillets are seasoned and prepared and the apples are made into my grandmother's skillet apple pies, a favorite that I've been making for over a decade. There is something inherently sweeter about a pie that is truly made by hand, especially one that is made with fruits from your own trees.

A Cottage on the Rolling Hills

Bluemont, Virginia

Set in the rolling hills of Virginia, this 1953 cottage makes a cozy little home with its vintage finds and well-considered decor. This bucolic dream belongs to Amy Whyte, an interior stylist who has made a name for herself with her thoughtful designs and her eye for antiques. The house sits on a two-acre plot, in a peaceful woodland setting in the Shenandoah Mountains. The view from the back deck is a little grassy meadow where you can watch deer grazing in the summer. In the fall, there are lovely views of the woodland gardens, planted with boxwood, lilacs, holly, and crepe myrtle by the previous owners.

"The first time I saw this house I fell in love with it," shares Amy. "We would drive by it on the way to the river with the dogs and I always thought it looked like something from a storybook … a little cottage in the English countryside. The day we saw the 'For Sale' sign in the front yard we called the realtor. We got in the next day and as soon I stepped in the door I knew. This was it! The realtor thought I was crazy."

The house hasn't been easy to renovate, but Amy says she could see the potential and charm within its bones. "It was built by Martha and Jack Hayes in 1953 as a weekend retreat for their family; Martha moved into the house full time after Jack passed and lived there by herself for years before having to sell when the rustic location got too much for her," explains Amy. "When I heard that, it kind of broke my heart. Walking through the house it was clear to me that Martha and Jack loved the house very much—Jack actually built

"THE FIRST TIME I SAW THIS HOUSE
I FELL IN LOVE WITH IT."
——— *Amy Whyte*

Chris Casson Madden
A ROOM OF HER OWN
SEASIDE Style
FAMILY HOUSES by the SEA

most of it himself, including some of the furniture and stained-glass mosaics. There were thoughtful personal touches everywhere. Their initials are even carved into the oak tree out front … I wrote a letter to Martha and submitted it with our offer. I wanted her to know that, if she decided to sell the house to us, it would be going to someone who loved it as much as she did, and that it would be well cared for. We decided to name the house Hayes Cottage after Martha and Jack."

After renovations were complete, Martha and Jack's daughter Suzanne reached out to Amy and shared a bit about her childhood growing up in the cottage. She wrote: "I can't even begin to tell you what a gift you've given me by restoring the cabin to the way it felt when I was a child, when we spent all our summers here in the Blue Ridge Mountains. We went to church at the Trappist monastery and would stop to pick bluebells and Queen Anne's lace along the river before heading to the farm stand to buy corn for dinner. It was magical." Suzanne also recalled spending the summer swimming and going blackberry picking with her friends. "My dad took us on hikes and taught me the names of all the wildflowers, such as wonderful Native American names like Pipsissewa. I imagine my dad put the heart on the oak tree. He was famous for his small love notes; there was an old school slate near the tiny kitchen on which he wrote 'I love you so …' It was indeed a house filled with love." And, thankfully, it still is.

A Timepiece of the Past

The Berkshires, Massachusetts

Western Massachusetts is home to fertile soils, rolling hills, a seemingly endless string of century-old small towns, and the Berkshire Mountains. For years people have flocked to this area, be it to set up a homestead, to leaf peep in the fall, or to enjoy the small-town life. Many come from nearby Boston, but for Susan Marsh the journey actually started in the Midwest, where she lived until her daughter Jess married a tenth-generation farmer from the fertile valley of Western Massachusetts. Susan and her husband soon felt the call to move from Michigan to Massachusetts themselves.

"When we moved to be close to our daughter, I was determined to find a special place where my desire for authentic surroundings could flourish," says Susan. The stars aligned and "a house, with a barn and charming chicken coop, set on a 28-acre wooded piece of heaven, presented itself like the best gift ever." The home's handcrafted windows, doors and hardware, along with its abundance of hand-planed wood walls, took their breath away. A walking tour of the rugged terrain was a magical "seal the deal" moment, as they feel most comfortable in quiet natural places where they say they are entertained by owls having conversations with one another, and an occasional visit by bears.

The home, a saltbox colonial-era style design, was actually built in the 1980s by the previous homeowner who wanted to create a timepiece of the past. Using designs from the era she worked to bring the house together with details and craftsmanship that have long been lost by more practical and less decorative design trends and methods of the past two centuries.

"The interior spaces of our house are a reflection of the serenity of the outdoors, its changing seasons, with a nod to historical interpretation," explains Susan. "A calm color palette along with lots of texture, including the aged patina of the wood, keep a balance of peacefulness while being interesting. The original owner rightfully boasts that there is not a single brushstroke of poly-anything on the wood in the house. The presence of imperfection fits perfectly with the handcrafted history and our desire to make things rather than buy them. There are hints of dog-claw graffiti and the most touching growth chart on an upright post, which we hope to carry on."

COZY(koh-zee)
adjective;

The home is one that feels timeless and comfortable, one that feels like living history, but still has its elements of modernity. With level floors, solar panels, and air conditioning throughout, it is a house that feels like it could live on for centuries to come.

An Historic Scottish Farmhouse

Ballston Spa, New York

Growing up in upstate New York, Richard Lanne Jr. dreamed of living off the land. His childhood was full of adventures he recorded in many notebooks, documenting the fleeting knowledge of the land and what he learned from the many scoutmasters and old timers that passed through his world. He grew up in a world akin to the *Foxfire* books of old, spending much of his youth learning how to forage, trap, hunt, build, and forge his future.

In 2002 Richard and his wife Jennifer, a painter, moved to Ballston Spa, New York, to a historic farm property once owned by an exiled Scot named Agnus McDearmid. They were searching for a fixer-upper and the 1779 farmhouse was just what they were looking for. "So many historic homes in the area were gutted somewhere between 1970 and 1990 and completely remodeled," says Jennifer, "but this one had the original floors and low ceilings we wanted. We love the character of the place."

Today the house, Jennifer's studio in the equally old barn, and a blacksmith's forge used by Richard in a former smokehouse, all feel cozy and eclectic. The spaces have been filled with handmade pieces and early American antiques. Jennifer's extensive collections fill each of the rooms and cause visitors to slow down and study each room at length.

INTERNATIONAL

"SO MANY HISTORIC HOMES IN THE AREA WERE GUTTED SOMEWHERE BETWEEN 1970 AND 1990 AND COMPLETELY REMODELED ... BUT THIS ONE HAD THE ORIGINAL FLOORS AND LOW CEILINGS WE WANTED. WE LOVE THE CHARACTER OF THE PLACE."

——— *Jennifer Lanne*

WELCOME

Jennifer's studio, a rambling space spread across two floors of the old barn, feels like a mix of the very old, the old, and the new. A disco ball hangs perfectly at home next to salvaged early wood finds and paintings created by Jennifer.

Richard and Jennifer have done much of the work on the property themselves over the past two decades. "Richard can really make anything, and that came in handy when we bought this farm," says Jennifer. "Any-time we needed a bracket for a door he was able to forge it; he really does know how to make just about anything. And if he doesn't, he teaches himself."

The house stands as an authentic example of a classic American farmhouse, one that was built to last, and that will continue to live on for many more decades to come.

Welcome Home

Waltons Creek, Kentucky

In my house, overhead and out of sight, the charred rafters hiding just above the guest bedroom and in the tiny attic tell a story of a long life well lived, one that in many ways has been charmed by rural farmhouse standards—especially in coal country, where the land is stripped of its minerals and anything in the path of Mr. Peabody's coal train, as the song goes.

This story-and-a-half farmhouse was built by my great-great-grandparents, it seems, but that is frequently up for debate. We do know that in 1900 the original house that stood here on this little hill in the middle of the valley burned down, leaving nothing but endless chunks of broken pottery and ironstone to be dug up each year when the garden is readied for spring. The house was rebuilt as a classic farmhouse, with five rooms on the first floor, all opening on to each other; a small central hall; and an upstairs that over time would be finished to become the bedrooms. A bathroom was added under the stairs when indoor plumbing came into trend.

The doors, likely original to the house, carry decades of wear—the bottom of the living room door marked with what seems to have been a red crayon in an impromptu canvas for a small child, most likely me, nearly 30 years ago. As told by the layers upon layers of wallpaper (my grandmother seemingly redecorated every few years), the house has evolved over the years and rooms have changed purpose: the upstairs bedrooms were once storage rooms for hanging onions and potatoes in.

K
TRUE PREP

After I inherited the house I wanted that evolution to continue. Aunt Peggy insisted I should not make this house a museum to the family, but instead fill it with the same sort of life that it has had since it was built over 120 years ago. I grew up with stories of my great-grandfather and his many businesses; of my great-grandmother making bread pudding for the farm workers with the left-over breakfast biscuits; of my dad's tobacco growing, and how he build the barns and garages with my grandfather by hand. Photos of family gatherings fill the old albums I used to pore over as a kid, stacked neatly next to a collection of my dad's childhood puzzles.

Today Aunt Peggy still tells of stories of her and Aunt Mary Lou playing with paper dolls on the steps and dancing on the front porch. She tells of having to grow up fast and help with the farming and housework after both of her parents died young. When asked which room of this house my great-grandparents died in she refused to say, instead telling me to focus on the good of the house and to make more good—a theme I try to carry with me as I continue to evolve the house and property.

Over the past few years I have slowly made the house my own, while continually referencing the past generations of the house. Loud curtains hang in what is now my dining room, but which for the first 127 years was the informal living room; the patterns of lilac and peonies remind me of my grandmother's flowers while the loudness of the fabric is a nod to her bold floral wallpapers and linoleum. The mantle in the room is original to the house, but was stored in the barn since the 1940s. Today it is back where it belongs, even if the house no longer has fireplaces.

A SIMPLE FEAST
BECOMING
Charlotte's Web

NEW WHEN I MET
GRAND ADVENTURE
AS ABOUT TO HAPPE
A.A. Milne

DREAM DECOR
A Wilder Life

The dining room is filled with collections from both sides of my family: a trunk I found in my Aunt Martha's garage, a washstand I found as a five-year-old in a tobacco barn my dad was having torn down; a pie safe my step-grandmother's first husband was restoring when he died. The latter was found in pieces in his long-abandoned shed and stapled and glued together to become what is probably my favorite piece of furniture in my house. The cabinet is filled with my grandmother's collection of 1980s Pfaltzgraff dishes, that don't go for much at yard sales but make me so happy to have.

My laundry room, once Aunt June's kitchen, before she built her own house next door, has become a favorite space in the house. A fan of white paint, I had refused to paint any room anything but white, until ripping off paneling in that room revealed Aunt June's old green wallpaper underneath. The walls, ceiling, and trim were painted to coordinate with the darkest colors in the wallpaper, a clear coat added to preserve the antique wallpaper, and the room filled with small pieces of the past, including toys of my dad's, a lamp of my grandmother's, a print from Aunt June's house next door, a dining table from my other grandparents' house, and other little knick-knacks that tell a story of the past.

When I was five or six, after learning that I would one day inherit this house, I went through with a crayon, marking all the pieces I wanted to keep. Today those pieces are spread throughout the house, paired with folk art and art I've collected from friends. My goal was to create a house that felt timeless and warm, one that you could always gather in be it for bread pudding in the dining room or a viewing of *Steel Magnolias*, on VHS, in the living room.

Today, after multiple fires, a tornado, my childhood painting adventures, and a century of living, the house still stands tall here in the valley. I know it is mine for only a generation, and then will, as it has for the past four, continue on after I am gone.

SUMMER
HENRIK IBSEN
267

Roadside Stands

As a child I had my pumpkin and watermelon stand each year. It sat in the yard down by the little country road we lived on. Each summer and fall I would plant, grow, pick, arrange, and sell dozens of varieties of pumpkin, while my grandfather grew the watermelons. I'd plan my displays months in advance, and each day we'd add freshly picked produce. I'd handpaint my signs and have them down by the main highway that ran down past the little country road. An honor jar or fish bowl would go out on a wrought-iron table and community members would come and get yellow, orange and red watermelons, or a whole slew of pumpkins and gourds. This tradition ended once I went to college, but my affinity for roadside stands lives on.

A FEW TIPS:

- Go early in the day before the summer heat sets in.
- Be flexible with your recipes; prepare what is in season.
- Bring cash. Most stands are cash only.
- Get to know your farmers. For many, this is their profession and they take great pride in what they produce and how they grow it.
- Spread the word. Roadside stands are the perfect photo moments. Share snaps on social media. Word-of-mouth marketing is a timeless tool.

Korey & Michael's Roadstand

Kentucky

Farming is something that just gets in your blood. No matter what you're farming, there is a drive to grow, to utilize the land, to put in the labor—the never-ending, back-breaking labor that brings forth the fruit. For Korey and Michael Brown, a pair of farmers in my community, the journey began as children and has grown into their own successful produce farm and ever-growing roadside stand.

In their late twenties, Korey and Michael had a dairy farm that simply didn't have a run of good luck. Korey says the day they started producing milk, the milk prices bottomed out. After four short years, the reality of a failing market brought an end to the dairy farm dream. The couple decided it was time to sell, but kept 25 acres in the hope of turning them into a successful business.

"Farming was always our way of life; Michael could do about every aspect of farming ... but produce was a new ballgame," Korey says of their transition from being dairy farmers to produce farmers. "We wanted a way to support our family that would allow us to work and grow together; we loved the idea of our children knowing how to work and interact with the community, as we believe the best life lessons happen on the farm." She continues, "We were told there was no way to make a living selling produce in a poor, rural community. But this community supported us when we went through cancer with no insurance; they saw us start from a card table on the side of the road; and when we grew, they cheered for us!"

The community certainly did support the ever-growing farmstead. Driving an extra five minutes out of town and into the county to buy from a local family was an issue to very few, especially if it meant not having to buy shipped-in produce from the big-box store in town. The goal isn't to get rich from being produce farmers; the goal is to always be able to put back one year's profits into the next year.

"Every year we try to add something new—a reason for people to come and check us out," explains Korey. "They come as customers but leave as friends. We added my dad's, Gray's Angus Farms, registered Black Angus meat at our stand and pork from the next county over." In recent years pick-your-own strawberries and flower patches have also been added. After a tornado hit the farm in 2021, Korey and Michael were able, as part of rebuilding, to add greenhouses to sell a variety of hanging baskets and bedding plants.

A mainstay for the farmstead, since the very beginning, has been a variety of pumpkins each fall. "When we pick pumpkins we always look for firm flesh, firm stems, and leaves around them dying," says Korey. "We always say they play hide and seek until they are ready to pick!"

Korey, a former teacher, says the next push is into agritourism, an ever-growing industry for farms looking to diversify and to provide education to the public. "I can't imagine not living in a place my children can run wild and explore. They have lived a life without borders. They have been able to let their imagination take them to so many places in the wild. Now that my children are grown, I can see the difference farm life made on them as well. I would put my children's work ethics up against any grown man's."

A Leaf Peeping Photo Walk

Kentucky

A slow cascade of color, starting far in the north, well beyond the boundaries of the nation, and well into the tiny islands of Newfoundland, works its way down toward the former colonies, first in the upper peaks of Vermont, New Hampshire, and the inland wilds of Maine, on into the Berkshires, the tiny towns of Massachusetts and Connecticut, into the quiet depths of the Catskills and Hudson Valley, marching ever south into the Appalachian Mountains and on to its final end, in early November, in the rolling hills of Natchez, Mississippi.

Leaf peeping is a time-honored annual tradition in the eastern United States that captivates everyone who respects and is amazed by the beauty of nature. Yes, leaf peeping is a real term, describing the activity of observing the changing colors of the leaves in the fall.

Fall in the east begins in early September and truly continues on through mid-November. The first signs will be the drying of the grasses and ground cover, and, of course, the vibrant colors of the goldenrod. The leaves will begin to change, with subtle hints of yellow appearing. An occasional red tree will appear. Slowly over the next two months the landscapes will morph into a full painterly scene of rich golds, oranges, reds, and yellows. The final phase of fall will be soft yellows and deep burgundies of the last remaining trees. In my opinion this final phase is the best time to view Central Park, while the earlier phases are best spent in the wilds of New England.

The autumnal spectacle absolutely captivates me each year. And much like with bird-watching, a camera makes the experience all the better. I personally prefer photographing on overcast days. I find the colors to pop much more when the contrast between light and dark isn't as strong. On sunny days it is too easy for the sky to become overexposed while you try to perfectly capture the color of the leaves. The harshness of a bright sun will also cause the camera to draw more contrast to the negative and positive spaces in the wooded areas, especially if the trees are in uniform parallel lines. An overcast day will have a softening effect that will make wooded images much more pleasing to the eye.

The differences between morning light and afternoon light will be distinct, but will make for better photos than midday sunlight. With morning light from the east you will have a light-box effect that will illuminate uniformly from one direction and allow for shadows to the west. Morning light will be cooler while afternoon light will have heavier, golden and yellow colors.

Use leaf peeping as an excuse to learn the various types of trees you are surrounded by. An arborist is a brilliant person to have with you, and they will encourage you to learn the various factors that reveal the type of tree you are observing. Leaf patterns and bark patterns are oftentimes the easiest. As you become proficient in identifying tree varieties, you will notice certain species change at certain times in the season. With this knowledge in mind you can better plan future routes to maximize the peak season for varieties of trees and regions of landscapes.

The most important thing to remember is to always have the camera ready. What might otherwise be mundane moments can become an interesting oasis of inspiration when the colors are at their peak. Don't be afraid to pull off the road (where it is safe) and take tons of shots. It is also important to be good at editing: you can bring out the color in the editing process.

In Fall

The garden slows down
as the wildflowers and weeds take over the once-pristine rows and furrows. Morning glories climb high into the spent cornstalks, and dried beans, long forgotten from picking dry in the late afternoon sun. The garden has a fresh planting of turnips and greens as we prepare for the heartier meals ahead. Tomato vines spread beyond their rows and occasionally still bear fruit. The garden has a certain beauty to it—a wildness.

Pumpkins are ready to be picked and used for decoration. Gourds, future birdhouses and indoor decor are picked and allowed to dry. The Indian corn is finally picked and used to festoon a fodder shock. The cellar is readied for the pumpkins that will soon be stacked along its shelves. Soup will be made from the pumpkins long into the winter months. A savory favorite made from freshly cooked pumpkin and basil from the garden is a fall staple here at the house. Any sort of pumpkin will do, but the best are the thick-fleshed heirloom kinds. The vines wither as the fruit turns a variety of shades of orange, or blue, or white, depending on the varieties. You may be tempted to pick your pumpkins early to start your decorating with, but the fruits will be best if you leave them until the vines have dried and died. Cut the stems with a pair of clippers or a swift, sharp knife. Never pull or break the stem, this will encourage the stem to become detached from the fruit and will lead to early rot.

The fall garden is largely one based on root vegetables, ones that can take the hot dry days and cooler nights. As October sets in, the threat of frost begins to loom and the garden is picked regularly. It is best to clear away the weeds and grasses regularly before they have a chance to head out and drop seed. This will make weeding much more difficult the following year. To keep the garden clean, clear the spent vegetable plants as they finish producing. I find a beauty in the untamed weeds and wildflowers of the fall in the garden, but this is something I must pay for the following year.

As you prepare for the end of the gardening season, take time to make notes of what grew and produced well and take note of the areas of the garden prone to be wet or dry throughout the year. We have found that my upper garden, the area farthest from my house, works best for the early-season crops. My lower garden, an area that spends much of the spring and early summer wet, is where traditionally large plots of turnips and greens have gone, sewn with a heavy hand rather than in narrow rows.

Dried Naturals

DIY

Fall is my favorite season to decorate. As a child I loved going through the seed catalogs each year to select which exotic varieties of pumpkins I would grow on the farm. Every year it became a thrill to decorate as much as possible with these pumpkins, the scarecrows, and the fodder shocks around the farm. As I've grown older I've begun to bring these outdoor elements inside more and more. I love the way dried grasses, grapevines, gourds, and flowers complement my collections of folk art and antiques throughout the season.

When planning your harvest-themed decor, go into the fields and fencerows, creeks, and even your yard. Never underestimate the value of a large grouping of dried hydrangeas or dried herbs. Goldenrod, the state flower of Kentucky, is my very favorite fall flower. Cut them early in the season, when they have just reached their peak gold color and dry them upside down in a dark space to ensure the colors don't turn to brown.

Inexpensive grapevine wreaths can easily be unwound and turned into bases for fall garlands. Fill with dried flowers, pinecones, nuts, or other elements that can be wired in. I personally love filling vintage jars and bottles with beautiful dried cattails, foxtails, and ornamental grasses for tall centerpieces that go vertical rather than horizontal. Naturally, never forget pumpkins. I still love the blue Jarrahdale pumpkins as much today as I did when I found my first one nearly 20 years ago. To finish your look, mix in candles in amber glasses, use flannel or wool blankets as table covers, and use tarnished silvers and brass elements.

RECIPE

Cast-iron Skillet Apple Pie

INGREDIENTS

½ cup (110g) butter
1 cup (210g) brown sugar
2 pie crusts
4 cups (480g) Granny Smith apples, peeled and sliced
½ cup (100g) sugar
cinnamon and nutmeg, to taste

It absolutely isn't fall until the first apple pie is made. When you are baking, the house even smells of the season. This is truly a simple dessert that will always please.

1. Preheat oven to 350°F (180°C).
2. While preheating, add butter and brown sugar to a 9-inch (23cm) cast-iron skillet and place it in oven to heat. Remove once the butter has melted.
3. Place one of the pie crusts in the skillet.
4. In a bowl, mix together the apples, sugar, and spices. Fill the crust with this mix and top with the second crust.
5. Bake for 20–25 minutes and serve warm.

Pea-picking Cake

CAKE

1 box (432g) yellow cake mix
4 eggs
½ cup (120ml) oil
1 11oz (310g) can mandarins

ICING

1 15oz (425g) can crushed pineapple
1 box Jell-O instant vanilla pudding mix
1 tsp vanilla
4 cups (940ml) whipped cream

This is my very favorite cake. When I was a child, Aunt June would make this cake every summer. It has a freshness that is perfect for the summer heat. Over the years it has become that cake that my friends request most often. I love a cake that starts as a box-cake mix. This allows you to whip up a cake at any moment. My pantry is always stocked with a dozen yellow cake mixes and the ingredients to easily pull together a variety of cakes at a moment's notice.

1. Beat all ingredients for three minutes and pour the mix into three well-greased eight-inch (20cm) cake rounds and bake at 350°F (180°C) for 30–35 minutes. Let the cake cool completely.

2. For the icing, mix the pineapple (along with its juice), pudding mix, and vanilla together. Fold in the whipped cream.

3. Ice the well-cooled cake. Best served cold.

GRAND TETON 1517
WARWICK 875
LEXINGTON 159
YOSEMITE 2201
WEST VILLAGE 860
SCOTTSVILLE 75
ACADIA 1355
MALIBU 2126
BEAVER DAM 9
ICELAND 3195

Winter

"The grass ain't greener,
Where it doesn't grow."
Greetings from

Winter on the farm

is quiet, still and calm. The leaves are gone, the crops are gone, and the fields are bare. A silence blankets the land. This is the season for resting and nesting. Rather than the garden, the porches, and the yard, the house is finally at the center of domestic life. A crisp air rattles against the west-facing windows and the house fills with light in place of the shade once created by the endless maple leaves.

The house feels cold even when it is warm as the winter wind creeps in through the cracks in the hundred-year-old Eastlake door and the slightly uneven upstairs windows. Spaces once reserved for the winter practice of tobacco stripping are now filled with stored lawn furniture, multiple lawnmowers, and a still-unhung basketball goal. Neighbors visit, but less frequently, as the cold tends to keep everyone inside. Meals made from pumpkins and root vegetables are served, and bread is baked with reckless abandon.

Christmas is a special season at the farmhouse. The house is filled with natural decorations, likely inspired by a childhood love of *Little House on the Prairie*. Early each December a Christmas tree is cut from a local farm and brought into the living room. The decorations are dried hydrangea, grapevine, and goldenrod, all collected from the yard and fencerows during the previous season. Dried gourds are added to fill in holes. The entry doors are lined with garlands made of freshly cut pine, blue spruce, and cedar. Sprigs of fresh greens and antique glass balls are tucked in every empty vessel and vase. The house feels especially warm during the Christmas season in a very old-fashioned way, with antique Christmas wares on display.

A Perfect Blank Slate

Weston, Connecticut

Not far from the hustle and bustle of New York City, in Weston, Connecticut, is a 1819 home with an unusual history. Now belonging to James Cregan and Michael Amigron, it was built over 200 years ago in a town that was later destroyed by the construction of the Saugatuck Reservoir in the 1930s. Before their town was flooded, a few fortunate homeowners were able to move their homes, including this house, which was moved a mile up the road.

It is now on a hill, set back from the main road—something James and Michael found uniquely special, as homes of this age are typically within feet of the road to allow for easier horse-and-carriage access in regions that historically have had heavy snowfall. The pair especially loves the hilly and rocky nature of the property, which feels like it is deep in the woods, but is just a short distance from nearby Weston. They knew they had to have this house as soon as they saw it. They were looking for a house that was built before 1900, was within commuting distance of New York City, and still retained the original architectural details such wide floors and beams—this fitted the bill perfectly.

James says the house felt like the perfect blank slate for a series of thoughtful renovations, designed to bring the house back to a period-appropriate decor and feel. Today the home is very much inspired by early American colonial architecture and furniture design but has charming touches of English countryside manors and cottages, with a subtle nod to their love of English gardens. "We are greatly inspired by color; we often see a color that we fall in love with and then design a space around that color," says James. "Also we're naturally inspired by antiquing and the special pieces that we find, as well as quality craftsmanship."

Encyclopedia
LAUREN
COLLECTION
HENRI MATISSE
HOWARD HODGKIN
HENRI MATISSE CUT-OUTS

CALDERONE COOK BEAUTIFUL
LEBOVITZ My Paris Kitchen
THE KINFOLK TABLE

"WE ARE GREATLY INSPIRED BY COLOR; WE OFTEN SEE A COLOR THAT WE FALL IN LOVE WITH AND THEN DESIGN A SPACE AROUND THAT COLOR."

—— *James Cregan*

A Fascinating Treasure Trove

Warwick, New York

In 2002 Jen and Micah O'Connor decided to head out of New York City, a place Jen had always called home, and move to Micah's hometown in the foothills of the Catskill Mountains. An idyllic pre-Revolution town, Warwick seems straight out of a storybook with its bustling downtown, thriving agricultural community, and rich history. Located in the Hudson Valley, about an hour north of New York, it's also ideal for people ready to move to the country while still wanting easy access to the big city.

The move has been such a success that over the years many of Jen's family members have also made their way up from the city and now live in or near Warwick. The O'Connors and their three children have embraced small-town life, taking part in local parades, festivals, and homecoming celebrations, as well as transforming their cookie-cutter suburban 1990s build into a fascinating treasure trove filled with great finds.

An urban planner-turned-art dealer, writer, and business consultant, Jen has raised three children while running a business from the dining room table for the past two decades. She's also found time to build an extensive collection of female-made art and folk art, as well as antiques and relics that would easily be enough to fill a museum.

Jen inherited her passion for collecting and Americana from her mother, Rosie, a well-known antique dealer in New York City. "Five-year-old me asked my mother to buy me a tea set at a church rummage sale," she recalls. "I wanted a real set for my house one day because my dolly set just would not do. While most mothers would have brushed away the request, most kids would not have made one, so you can imagine what happened."

"It's not the individual things that are important to me—it's the moments and memories they keep present that resonate and establish value for me," she explains. "As a keeper of old, handmade and collected things, I am often reminded to be grateful where I am, with a nod to how I got here. When an object shows its age from use it's like a wink to the worth of wisdom that comes only with experience."

Jen has made it a subtle point to only have original works throughout her house, and many of the most fascinating artworks in the house have been created by friends of the Connors. Each piece comes with a story of who created the work and how their paths intertwine with the owners of this house. "Collections should be deeply personal; collected objects should not be about value, but about harnessing sentiments and celebrating the tiny accumulations of bits of magic over time," she says. "I've collected Torquay art pottery made before World War I since I was 18, so in 35 years of hunting for this elusive British ceramic art it has accumulated across shelves and filled cabinets across the decades."

"WHEN AN OBJECT SHOWS ITS AGE FROM USE IT'S LIKE A WINK TO THE WORTH OF WISDOM THAT COMES ONLY WITH EXPERIENCE."

Jen O'Connor

A Joyful

Her collections are all the more impressive when you realize that most objects were real finds: "I don't go on a mission to buy, I let things come to me in my travels and adventures," Jen explains. "I travel lightly so I have room to bring home the irresistible in the suitcases. I can never pass up anything that looks like it was well loved—a teddy bear, a worn velvet jewelry box, a favorite silver pitcher dented from constant use."

Jen displays her finds in "small towers of texture, built like tiny altars to good memories," displays which she loves to rearrange often to suit her mood. "There has never been a time when old things were not interesting to me," she concludes. "My home feels sometimes like I am living in a scrapbook of mementoes from adventures. I'm grateful that my husband absorbs my style and sentimentality as just part of me. Of course, he knew me when I hung my collection of hats on the wall of my NYU dorm next to a giant old painting that now hangs in our mudroom. I guess I blame that first tea set and my mom who loved old things too."

JAMES E. GRANEY
BROOKLYN
NEW YORK

Forget Me Not

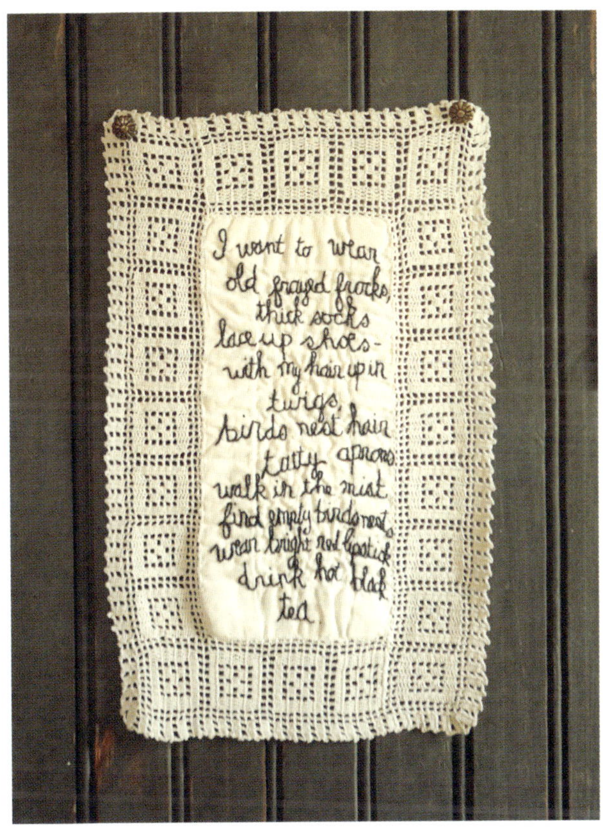
I want to wear
old frayed frocks,
thick socks
lace up shoes -
with my hair up in
twigs,
birds nest hair
tatty aprons
walk in the mist
find empty birds nests
wear bright red lipstick
drink hot black
tea

A 250-year-old Farmhouse

Chesterfield, New Jersey

Shannan and Drew Haupt's New Jersey home dates back to the 1740s, and has undergone plenty of additions and transformations during the past 250 years. As the couple have made this house their home, they have been acutely aware of the responsibility to make changes while respecting its long history. After all, they will be just a chapter in a home that is nearly one hundred years older than the nation they reside in.

High-school sweethearts Shannan and Drew grew up in North Carolina, where they both admired the early settlements of their hometown; the stories of the past captivated them. Settling into this old home fulfilled a lifelong dream of living in a historic house.

"My favorite pastime is antiquing and thrifting pieces that are unique and speak to my soul. Old things have better stories to tell," says Shannan. As the couple spent weekends browsing antique stores, they used it as an opportunity to explore New Jersey. On one trip, as they passed a rural community, surrounded by horse farms, and in many ways reminding them of North Carolina, they stumbled across the house they now call home. "I remember when we first visited this house it took my breath away," recalls Shannan. "Walking through each room pulled you in more than the previous and each original feature stole a piece of my heart that day."

"I love the 250-year-old wide plank yellow pine flooring, the exposed beams and floorboards that have settled and contoured over the centuries, and the three fireplaces." Humble in its origins, the home is very primitive in its construction; nothing is level. The couple is slowly un-renovating the modern updates that were done in the kitchen, laundry, and bathrooms during a renovation flip, and adding in more period-appropriate fixtures.

"We have restored a few original warming cupboards that were almost lost to time hidden behind the walls of modern updates. I found some old online listings with photos, where I saw a tiny hint of a cupboard door ajar in the main bedroom next to the fireplace. I wondered if it could still be there and had to find out. The other original cupboard was pictured in the main living room. Giving those original cupboards back to the home has been one of our favorite restoration projects, gaining back some irreplaceable history as well as restoring much needed storage space and charm back to the home."

"Although there is much left to do, our time spent in this house has already been so rewarding. With each change we try to give the house back some things that were lost over the years using our combined skillsets, resources, vision, and time."

"WITH EACH CHANGE WE TRY TO GIVE THE HOUSE BACK SOME THINGS THAT WERE LOST OVER THE YEARS USING OUR COMBINED SKILLSETS, RESOURCES, VISION, AND TIME."

——— *Shannon Haupt*

The Farm

Boston, Massachusetts

Just outside Boston, west of the streets that laid the foundation of a future nation, and west of the modern skyscrapers, and of suburban spread, is a farm that has been around for years. Charlescote Farm is a fourth-generation farm nestled into the Charles River valley of Massachusetts. It was originally a summer home for a Boston family, but has since become a year-round residence, and livestock and dairy farm. The farm, well-seasoned at this point, has seen a lot of life come and go.

In its earlier years Charlescote Farm had a flock of turkeys, was known for its herd of champion Ayrshire cattle, and even had a fox-breeding operation. The estate also served as a destination for those in the know, during Prohibition, offering spirits from the vault nestled into its fieldstone foundation.

What was once a serious dairy operation that supplied milk to HOOD, a regional dairy manufacturer, the farm now has Hereford cattle and serves as a beef farm, supplying local grass-fed beef to members of the community and beyond. The estate presently has beef cattle, a flock of sheep, more chickens than one can count, a few pigs seasonally, and crops. In current times it is as lively as ever with its animals, fruit trees, and constant activity.

9521

FRESH
EGGS

Greetings From
WALTON
CREEK

In Winter

The days grow short, the mornings are covered in crisp, sharp frost that refracts and reflects the bluish early to mid-morning light. The land is at rest. The season is one of nostalgia for me. The season feels like a time for Christmas, for the smell of hickory nut pies, for time spent indoors with the Saturday-morning cartoons, awaiting the fall of snow. The garden in this season is oftentimes an afterthought, but it is a time for ordering seeds, dreaming of spring, and clearing the remains of the previous season. On a good year my cellar will be filled with pumpkins, picked from my own vines in late September, used for fall decorating, and later put in the cellar to be saved for wintertime pumpkin pies and pumpkin soup, a dish I am frequently requested to make.

Planning well in the winter will ensure a well-filled larder and cellar the following winter. Use these days and hours spent inside to order seeds, to take notes on what you enjoyed about the previous garden, and what you dream of for the year head. Seeds are best ordered from a catalog, to stay true to nostalgia, and should be ordered early to ensure the desired varieties are not sold out. Use this time to read almanacs and books that were written by experts on developing the most wonderful and practical gardens. *Martha Stewart's Gardening* is the best resource I have found for planning a garden.

The winter is also a fun time to craft with dried gourds, okra pods, and corn shucks from the summer months. Allow your mind to wander back in time and create handmade objects. Keeping an active body and mind during the winter months will be beneficial for being prepared for the grueling spring planting ahead. As with everything in life, rest is required. The winter allows both garden and gardener to rest.

Christmas at Hort & Pott

Hudson Valley, New York

"I love the sensorial harmony and vibrancy of the Christmas season: the scent of fresh pine and spruce; the sounds of bells, carriages, and carols; the warm light of candles and fireplaces; the twinkling glow of Christmas tree lights; and, of course, the timeless tradition of bringing the outdoors in—the magical experience of Christmas truly is the embodiment of all the elements that we hold so dear at Hort & Pott," says designer and shop owner Todd Carr, who, along with Carter Bates, has created a magical world somewhere deep in the Hudson Valley.

The two exchanged a very busy life in New York City for a life in the wilds a few hours north. Known for their designs using natural elements and plants native to the Hudson Valley region, the pair really bring the festive season to life with their exuberant Christmas decorations.

"At home, we start our Christmas celebrations by searching for a tree at our favorite local farm. We'll hold a decorative ceremony by playing some of our

favorite music, like Phil Spector's 'A Christmas Gift For You,' light the fires, and embellish the tree with our favorite glass woodland ornaments of creatures, birds, pinecones, mushrooms, and more," says Todd, who has been collecting since he was a teenager. It creates a gleaming spectacle and woodland sanctuary to ward off winter colds and a welcoming space in which to cook hearty meals and relax around the table, while crafting new ornaments and treats.

At Hort & Pott, they take the idea of Christmas and run with it, crafting wreaths, garlands, and ornaments, and attempting to capture the nostalgic essence of the holiday through vignettes, photography, and botanical installations. It's a time to celebrate, and they want to inspire people to get into the spirit by hosting wreath-making and crafting workshops, and offering freshly foraged wreaths, garlands, swags, and homemade ornaments celebrating the natural world. They take decorating to the next level by incorporating a lush bounty of lights and botanicals foraged from the Catskills. Their favorite Christmas decoration to work with is the wreath: the eternal symbol of nature and a timeless classic that can be reinterpreted in many ways.

"The wreath perfectly encapsulate the season's greetings," explains Todd. "A wreath can be simple or complex, and may incorporate native, local botanicals as well as handcraft decorations." Other ways to bring natural elements indoors for Christmas include displaying some freshly cut evergreen branches on mantels and table centerpieces, or hanging them from a door or banister. "While we love berries, we tend to keep the 'messier' plant materials outdoors because they drop, requiring a clean-up effort. One of our favorite simple ideas is filling a bowl with pinecones, with sprigs of white pine peppered in for a festive and understated table piece that can still evoke the feeling and smell of the woods. It's really about capturing the essence of nature. And of course, we'd recommend keeping fresh-cut greens away from any heat source for longevity."

Todd and Carter draw their inspiration from a mental library of the holidays they've been developing for years, drawing on age-old customs and reworking and experimenting with ways of adapting and reinterpreting them. "Our inspiration has developed from looking at myriad customs, both old and new, from various parts of the world; we have a fondness for the old ways, when bringing the outdoors in wasn't just a decorative practice but a magical one to celebrate nature and her cycles."

Venice

Candlemaking

Trish Baden of candle brand Flores Lane may be based in the heart of West Hollywood, but her roots are deeply planted in the rural areas of the Midwest. Long before she moved from Ohio to Los Angeles, Trish was actively seeking ways to live a slower pace of life and to educate those around her on the importance of whole and clean living. With her candle business, Trish has made a commitment to using the most health conscious of ingredients and resources. She shares how to make your own candles at home, wherever home might be.

SUPPLIES

Vessel Mason jars are safe to experiment with because they can sustain high temperatures.

Wax Soy wax is natural, burns clean, and for twice as long as paraffin candles. Paraffin wax is a petroleum by-product created when crude oil is refined into gasoline. Beeswax is perfect for when you need to make pillar candles and want to stick to an all-natural blend; the only issues are that it's costly and its natural scent can interfere with fragrances.

Wicks Use wicks made of all natural cotton fiber and beeswax coating with an aluminum metal base. We use glue dots and wick sticks to keep the wicks in place while pouring.

Pitcher We highly recommend using a metal pitcher for safety and ease.

Oils You'll need essential oils and a carrier oil (we use grapeseed). You can also buy fragrance oils online, which are pre-blended and ready for candlemaking. Check for the oil's flash point (usually located on the bottle) to see at what temperature the fragrance should be added to the mixture, prior to pouring.

Mixing
Grill skewers work well.

Fragrance Picking a fragrance oil or blend for your candle can be difficult if you have no idea what types of scent you like. If you have a favorite essential oil you like, do not blend directly into candle wax! The real difference between essential oils and fragrance oils is that essential oils typically contain only plant oils, whereas fragrance oils have the essential oils blended with a carrier oil (like grapeseed or jojoba) to elongate the burn time of the scent.

SETTING UP

Make sure you have a flat, dry area that can be easily covered with kraft paper, plastic wrap, garbage bags, or newspaper. Tape down if needed.

1. Place the wax into the metal pitcher. Heat to 160°F–180°F (70–80°C).

2. While the wax is melting, prepare the jars by affixing the wick's aluminum tab to the bottom of the jar. You can use a glue dot or hot glue gun. Pro tip: make sure you space the wicks out evenly if you're using multiples to ensure they can breathe. The more stable the flame, the more even the burn!

3. Stabilize the wick with the stick (metal or wood). Cut off the end of the wick, leaving 1 inch (2.5cm) of wick above the stabilizer.

4. Once the wax is melted to liquid, let cool for a few minutes.

5. Add the fragrance, using 1–1.5 oz (30–45ml) of fragrance oil per 1 pound (450g) of wax. If you're using essential oils, make sure to dilute it with a carrier oil like grapeseed.

6. Pour the wax into the vessels. Let each cooling candle sit at least 2 or more inches (5cm) away from each other.

7. Now we wait!

   Sometimes a candle will crack or even get pockets of air that get caught inside, causing them to collapse. If your candle looks like cottage cheese, you've either added too much fragrance or overheated it. You can create a finished look by heat gunning the top or using a hair dryer that has a heated feature. Sometimes I pour a little layer of new wax on top to seal it in if there's space in the jar.

Caramel Bread Pudding

INGREDIENTS

1 loaf sourdough bread
3 tbsp butter, melted
½ cup raisins (optional)
3 eggs, beaten
2 cups (470ml) whole milk
½ cup (100g) brown sugar
½ cup (100g) white sugar
1 tsp ground cinnamon or pumpkin pie spice
1 tsp vanilla extract
Whipped cream or caramel sauce for topping

This timeless Southern classic can be as simple or as complex as you'd like. This is the basic recipe with great cinnamon and vanilla flavors but can be added to easily. The key is using a really good bread—I use sourdough.

1. Preheat oven to 350°F (180°C) degrees. Cube bread into 1-inch (2.5cm) cubes.

2. In a large bowl, mix all ingredients and whisk well. Make sure the eggs are well incorporated. Add the bread cubes to the mix and allow to soak up the mixture.

3. Turn out into a well-greased baking pan (I use an over-sized pie dish). Pour any remaining liquid over the dish. I sprinkle on additional spices for more flavor.

4. Bake for 45 minutes or until the liquid feels firm. Do not burn. You can cover with aluminum foil if desired, as it will prevent the top from becoming overly dry.

5. Once cool, top with whipped cream or caramel sauce and serve cold or hot.

   This recipe can easily be modified for additional flavors; bourbon is a common addition.

DESSERTS
NO. SERVINGS
NAME OF RECIPE
Italian Cream C
Ingredients
1 stick butter
1/2 cup oil
1 cup Coconut
2 cups flour
2 cups sugar
5 egg yolks
5 eg whites (stiffly beaten)
1 cup
1 cup
1 t. vanilla
Method
Cream butter and oil with sugar - add
one at a time - add 1/2 t. soda
milk - alter. with flour - fold in egg
vanilla - add floured nuts & coconut.
3 - 9" pans - floured & greased.
NO. SERVINGS
NAME OF RECIPE
Ingredients
Method

Italian Cream Cake

CAKE

½ cup (110g) butter
½ cup (120ml) oil
½ tsp baking soda
1 cup (100g) shredded coconut
2 cups (250g) all-purpose flour
2 cups (400g) sugar
5 egg yolks
5 egg whites (stiffly beaten)
1 cup (225ml) buttermilk
1 cup (125g) chopped pecans
1 tsp vanilla essence

ICING

1 cup (220g) butter
1 pound (450g) icing sugar
2 tsp vanilla
8 oz (230g) cream cheese
½ cup (50g) shredded coconut
¼ cup (40g) chopped pecans

As a child this was the fancy cake that would be prepared for church gatherings, funeral dinners, or the Ashby Christmas. This cake was one that intimidated me as a child and it took me many years before I plucked up the courage to master it. In reality, it is a very easy and delicious cake, sure to always be a crowd-pleaser, and one that feels instantly nostalgic.

1. Cream the butter and oil with the sugar. Add the egg yolks one at a time.
2. Add half a teaspoon of baking soda to the buttermilk, then add to the previous mixture, alternating with the flour.
3. Fold in the egg whites and vanilla. Add the pecans and coconut.
4. Pour into three 8-inch (20cm) greased cake rounds. Bake at 350°F (180°C) for 30 minutes.
5. For the icing, cream the butter and softened cream cheese. Add the sugar and vanilla. Fold in the nuts and coconut.
6. Ice the well-cooled cake.

Thank You

The small town life isn't for everyone, some people just belong in the cities, but for those that yearn for wide open spaces and single stoplight towns this book is for you. Out here in the middle of nowhere places, community is a cornerstone and an appreciation for being a part of a community is fundamental. Every day I am grateful for my small town roots. I am grateful to be able to live two doors down from where I grew up, surrounded by generations of people that believe in celebrating our small towns and small town lives. Ohio County, Kentucky, my hometown, is a pretty special place.

To Dru Ann for teaching me that we should never be limited by our small town roots, to Linda for teaching you're never too old to climb a tree, to Jean for teaching me the importance of a garden-grown turnip, and to Mrs. Bradley for teaching me the importance of the written word and shared story.

Texts: *Ben Ashby*
Photography: *Ben Ashby*
Design: *Elise Castrodale*
Editing: *Léa Teuscher*

@folkmagazine

Sign up for our newsletter with news about new and forthcoming publications on art, interior design, food & travel, photography and fashion as well as exclusive offers and events. If you have any questions or comments about the material in this book, please do not hesitate to contact our editorial team: art@lannoo.com

ISBN: 978 94 014 8211 0
D/2023/45/216 – NUR 450/653

www.lannoo.com